Farm Animals
Cattle

Heather C. Hudak

![W logo]

Weigl Publishers Inc.

Published by Weigl Publishers Inc.
350 5th Avenue, Suite 3304, PMB 6G
New York, NY 10118-0069
Website: www.weigl.com

Copyright © 2007 WEIGL PUBLISHERS INC.
All rights reserved. No part of this publication may be reproduced,
stored in a retrieval system, or transmitted in any form or by any means, electronic,
mechanical, photocopying, recording, or otherwise, without the prior written
permission of the publisher.

Library of Congress Cataloging-in-Publication Data

Hudak, Heather C., 1975-
  Cattle / Heather C. Hudak.
     p. cm. -- (Farm animals)
  Includes index.
  ISBN 1-59036-422-8 (hard cover : alk. paper) -- ISBN 1-59036-429-5 (soft cover :
alk. paper)
  1. Cattle--Juvenile literature. I. Title.
  SF197.5.H83 2007
  636.2--dc22
                              2005034667

Printed in the United States of America
1 2 3 4 5 6 7 8 9 0 10 09 08 07 06

Editor Frances Purslow
Design and Layout Terry Paulhus

Cover: There are nearly 1.5 billion cattle in the world today.

All of the Internet URLs given in the book were valid at the time of publication.
However, due to the dynamic nature of the Internet, some addresses may have
changed, or sites may have ceased to exist since publication. While the author and
publisher regret any inconvenience this may cause readers, no responsibility for any
such changes can be accepted by either the author or the publisher.

Every reasonable effort has been made to trace ownership and to obtain permission
to reprint copyright material. The publishers would be pleased to have any errors
or omissions brought to their attention so that they may be corrected in
subsequent printings.

Contents

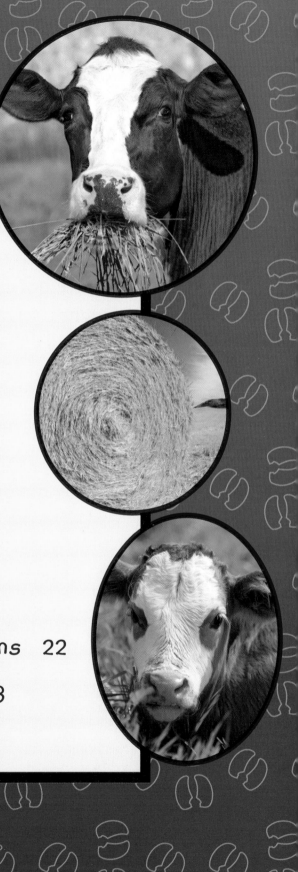

Meet the Cattle

Cattle are large farm animals. They have heavy bodies, long tails, and **cloven-hoofed** feet. Most cattle have a thick coat of short hair. A few **breeds** have long hair. Cattle come in many sizes, shapes, and colors.

Some cattle have horns. Their horns are hollow and have no branches. Cattle born without horns are called polled. Most farmers **dehorn** their cattle. This is usually done when baby cattle are 3 weeks old.

Cattle are **mammals**. They are also social animals. Cattle live in large groups called herds. They **graze** together in large, green fields called pastures.

Cattle must drink 2 gallons (8 liters) of fresh, clean water to make 1 gallon (4 L) of milk.

Most cattle do not like the taste of buttercups.

All about Cattle

Cattle are sensitive animals. They are easily frightened. Their natural reaction to danger is to flee. Cattle are also sensitive to high-pitched sounds. A human's voice is more alarming to them than any other sound.

There are three main groups of cattle. They are grouped by their use or purpose. The three groups are beef cattle, dairy cattle, and dual-purpose cattle. Beef cattle are raised for their meat, while dairy cattle are raised for their milk. Dual-purpose cattle are used for both meat and milk.

Cattle can form friendships. They show affection by grooming and licking one another.

Breeds of Cattle

Angus

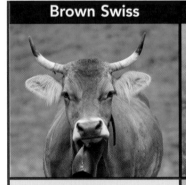

- Beef cattle from Scotland
- Hornless
- Black coat
- First arrived in the United States in 1873

Ayrshire

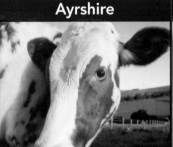

- Dairy cattle from Scotland
- Shades of red and white with spots
- Purebred

Brahman

- Beef cattle from India
- Have a large hump on their shoulders
- Gray, red, or light black

Brown Swiss

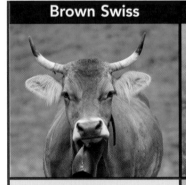

- Dairy cattle from Switzerland
- Color varies from chestnut to black
- Grayish stripe down the back

Hereford

- Beef cattle from England
- Brown coat on body, white markings on head and under belly

Jersey

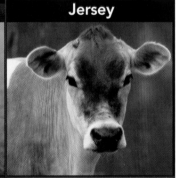

- Dairy cattle from England
- More tolerant of heat than larger breeds
- Color varies from light gray to creamy brown

Cattle History

POLAND

All cattle today come from aurochs, or wild oxen. At one time, aurochs lived throughout Europe, Africa, and Asia. They were first tamed about 8,500 years ago. The last aurochs lived in Poland in 1627.

In 1493, Spanish explorer Christopher Columbus brought long-horned cattle from Spain to Santo Domingo in the Dominican Republic. These cattle were later taken into Mexico and eventually into Texas. The first cattle in the United States were the Texas Longhorns.

Pilgrims from Great Britain brought cattle to New England in 1624. Cattle raising spread westward as **pioneers** moved across the country.

Cattle live in groups called herds. Most cattle will naturally follow the smartest and strongest members of their herds.

Ranchers have a cattle drive to move their herds to new pastures. They ride on horseback beside the cattle.

Cattle Shelter

Most cattle live on farms. They must be protected from poor weather. Large evergreen or oak trees provide shade from the Sun. Trees also break strong winds. A shed or barn provides shelter from rain and snow.

Pasture should be fenced to keep cattle in and **predators** out. Fences can be made of wood or wire.

Cattle become stressed in new pens. They also do not like to be alone and should be able to see their herd mates from their pens.

Cattle are very affectionate. They can become attached to their caretakers.

Cattle have very good memories. They can find their way back to their favorite pastures and watering holes.

Cattle Features

Cattle are **hardy** animals. Their bodies are **adapted** to their environment. They can live in hot and cold weather. The coat of hair on their body grows thicker in winter. Their muscular backs and legs allow them to graze all day. Other parts of their bodies have special features and uses.

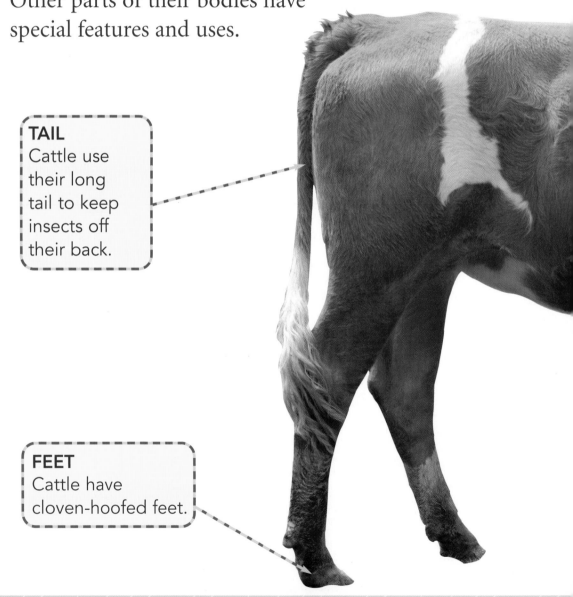

TAIL
Cattle use their long tail to keep insects off their back.

FEET
Cattle have cloven-hoofed feet.

EYES
Cattle's eyes are located on the sides of their head, so they can see all around them.

NOSE
Cattle have a good sense of smell. They can smell things as far as 6 miles (10 kilometers) away.

TEETH
Cattle have 32 teeth. They have no top front teeth.

What Do Cattle Eat?

Cattle are herbivores. Herbivores eat plants. Cattle eat grass, leaves, hay, shrubs, and grains. They can eat up to 95 pounds (43 kilograms) of feed each day.

Baby cattle eat whole, cracked, or rolled grains. It is also a good idea to feed cattle protein pellets. This is another source of **nutrients**. Cattle need to have fresh, clean water, too.

Cattle graze for 6 to 9 hours each day. They graze by wrapping their tongues around a plant and snipping it off with their lower teeth. Cattle eat the top of the plant before eating the stems. They prefer young plants because they are tender and easier to eat.

Cattle are ruminants. Ruminants have many parts to their stomach. Cattle have a four-part stomach to help them break down the plants they eat.

There are more than 800 million acres (3.2 million sq. km) of grazing land in the United States.

Cattle Life Cycle

Young female cattle are called heifers. Once they become mothers, they are called cows. Male cattle are called bulls. Baby cattle are called calves.

Like humans, cows **gestate** for about nine months. Before giving birth, they swish their tails and wander off.

Newborn

At birth, calves weigh about 80 to 100 pounds (36 to 45 kg). They stand up on their wobbly legs soon after birth. Calves suckle milk from their mothers. Cows will continue to give milk for about 300 days. Cows and calves stay close together to form a strong bond.

3 Months to 2 Years Old

Calves stop drinking their mother's milk when they are 3 or 4 months old. At this time, they begin to graze with older animals. Cows can have babies when they are 2 years old. At this age, cows weigh about 1,100 pounds (500 kg).

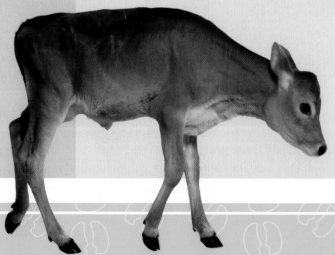

When cows give birth, it is called calving. If calving takes longer than 1 to 2 hours, problems may arise.

Cows often give birth to their calves in the spring. Calves grow quickly. With special care, cattle can live up to 25 years.

Adult

Cattle stop growing at 4 years of age. Most adult cattle are between 57 inches and 66 inches (1.5 and 1.7 meters) tall. They weigh between 800 and 2,500 pounds (362 and 1134 kg).

Caring for Cattle

Cattle need to visit the **veterinarian** once a year. They should be **vaccinated** against certain **diseases**.

Some plants are poisonous to cattle. These plants include yew, hemlock, and bracken. They sometimes grow in gardens or fields where cattle graze. If cattle eat these plants, they may become ill or die. Sometimes, cattle get **parasites**. These include lungworms, ticks, and lice. Cattle with parasites need special treatment.

Cattle often have hoof problems. Sharp stones, pieces of glass, or nails can get caught in their feet. Very warm weather causes heat stress. Cattle with heat stress pant and drink large amounts of cool water.

Useful Websites

To learn more about cattle and other farm animals, visit: **www.enchantedlearning.com**. Type "cows" into the search box and click "Search."

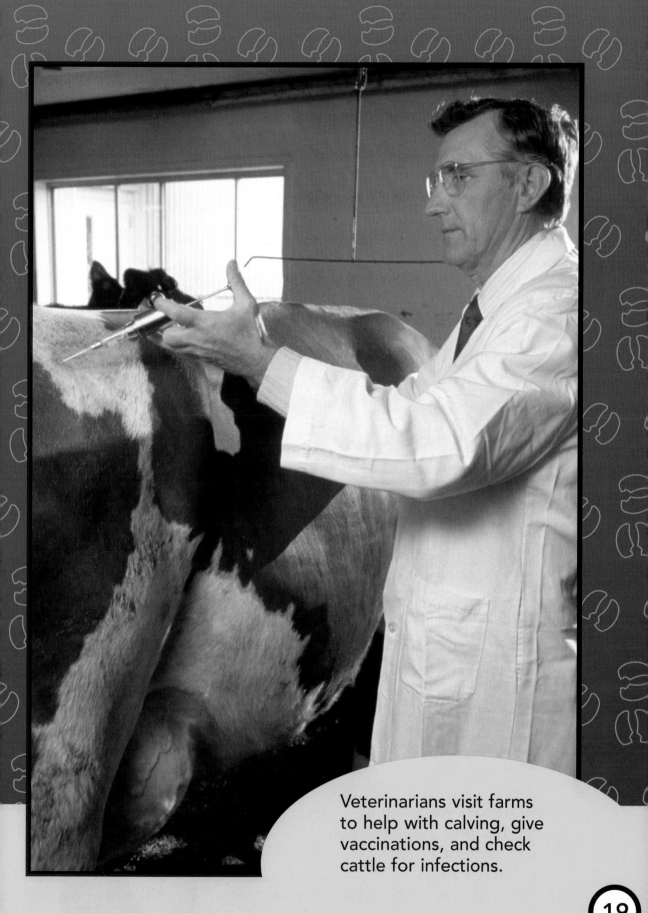

Veterinarians visit farms to help with calving, give vaccinations, and check cattle for infections.

Myths and Legends

Cattle are treasured animals to people around the world. For centuries, people have shared stories about cattle.

In ancient Egypt, the goddess Hathor was often shown in the form of a cow. Egyptians believed that the Milky Way was made of milk that flowed from her udders.

Hindus believe in a **sacred** cow named Kamadhenu. She is the mother of all cows. Hindus will not harm cows in any way. In Hindu countries, such as India and Nepal, cows roam the streets freely. On special days, they are decorated with flowers, paint, and ornaments.

Cow statues were found in the tomb of Tutankhamun, one of ancient Egypt's best-known kings.

Ice and Fire

There is a Norwegian legend that the world began as two parts. One part was made of ice, and the other was made of fire.

One day, fire melted some ice, making a puddle of water. Two creatures came from this puddle. They were a giant named Ymir and a cow named Audumla.

Audumla lived in the icy land. She ate snow and salt. Audumla made milk for Ymir to drink.

Audumla licked a block of ice. After licking for three days, she could see a man inside the ice. She freed him. His name was Buri. He was the first god. Buri made many other gods. Audumla and Buri used Ymir's body to make Earth and the sky. The gods made people from Ymir's eyebrow.

Frequently Asked Questions

Why do humans drink cows' milk?

Answer: Milk is very healthy for humans. It contains calcium and vitamin D. These nutrients make strong bones and teeth.

What is mad cow disease?

Answer: Mad cow disease harms the brain of cattle. It can cause the animal to become very ill and die. The disease spreads when cattle eat the meat of another animal that had mad cow disease.

How do I keep track of the cattle in my herd?

Answer: You can use ear tags to keep track of your cattle. Ear tags give information about cattle. They can tell where a cow or bull came from, where it has been, and what other animals it has been near.

Puzzler

See how much you know about cattle.

1. Name three main groups of cattle.
2. Where did cattle first come from?
3. How many hours do cattle graze in a day?
4. How many months do cows carry their babies before giving birth?
5. What is the name of baby cattle?

Answers: 1. Dairy, beef, and dual-purpose 2. Europe, Africa, and Asia 3. 6 to 9 hours 4. 9 months 5. Calves

Find Out More

There are many more interesting facts to learn about cattle. If you would like to learn more, take a look at these books.

Malachy, Doyle and Angelo Rinaldi. *Cow.* New York, NY: Margaret K. McElderry, 2002.

Rath, Sara. *The Complete Cow.* Stillwater, MN: Voyageur Press, 2003.

Words to Know

adapted: adjusted to the natural environment

breeds: groups of animals that have common features

cloven-hoofed: feet that are divided into two parts

dehorn: to surgically remove an animal's horns

diseases: illnesses

gestate: to carry offspring in the womb

graze: to eat grass in a field

hardy: able to survive in a harsh environment

mammals: animals that have warm blood and feed milk to their young

nutrients: parts of food that nourish living things

parasites: living things that feed off and live on other living things

pilgrims: people who journeyed to the United States in 1620

pioneers: the first people to settle a new territory

predators: animals that hunt other animals for food

sacred: highly respected

vaccinated: given medicine to prevent diseases

veterinarian: an animal doctor

Index